THE TAU MANIFESTO

Michael Hartl

Tau Day, 2010
updated Tau Day, 2022

1 The circle constant

The Tau Manifesto is dedicated to one of the most important numbers in mathematics, perhaps *the* most important: the *circle constant* relating the circumference of a circle to its linear dimension. For millennia, the circle has been considered the most perfect of shapes, and the circle constant captures the geometry of the circle in a single number. Of course, the traditional choice for the circle constant is π (pi)—but, as mathematician Bob Palais notes in his delightful article "π Is Wrong!",[1] π *is wrong*. It's time to set things right.

1.1 An immodest proposal

We begin repairing the damage wrought by π by first understanding the notorious number itself. The traditional definition for the circle constant sets π equal to the ratio of a circle's circumference (length) to its diameter (width):[2]

$$\pi \equiv \frac{C}{D} = 3.14159265\ldots \tag{1}$$

The number π has many remarkable properties—among other things, it is *irrational* and indeed *transcendental*—and its presence in mathematical formulas is widespread.

It should be obvious that π is not "wrong" in the sense of being factually incorrect; the number π is perfectly well-defined, and it has all the properties normally ascribed to it by mathematicians. When we say that "π is wrong", we mean that π *is a confusing and unnatural choice for the circle constant*. In particular, a circle is defined as the set of points a fixed distance, the *radius*, from a given point, the *center* (Figure 1). While there are infinitely many shapes with constant width (Figure 2),[3] there is only one shape with constant radius. This suggests

---

[1] Palais, Robert. "π Is Wrong!", *The Mathematical Intelligencer*, Volume 23, Number 3, 2001, pp. 7–8. Many of the arguments in *The Tau Manifesto* are based on or are inspired by "π Is Wrong!". It is available online at https://www.math.utah.-edu/~palais/pi.html.

[2] The symbol $\equiv$ means "is defined as".

[3] Image retrieved from Wikimedia on 2019-03-12. Copyright © 2016 by Ruleroll

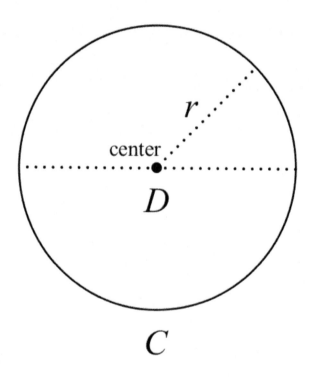

Figure 1: Anatomy of a circle.

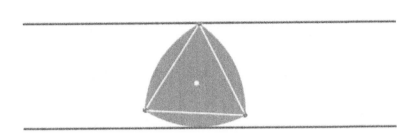

Figure 2: One of the infinitely many non-circular shapes with constant width.

that a more natural definition for the circle constant might use r in place of D:

$$\text{circle constant} \equiv \frac{C}{r}. \tag{2}$$

Because the diameter of a circle is twice its radius, this number is numerically equal to 2π. Like π, it is transcendental and hence irrational, and (as we'll see in Section 2) its use in mathematics is similarly widespread.

In "π Is Wrong!", Bob Palais argues persuasively in favor of the second of these two definitions for the circle constant, and in my view he deserves principal credit for identifying this issue and bringing it to a broad audience. He calls the true circle constant "one turn", and he also introduces a new symbol to represent it (Figure 3). As we'll see, the description is prescient, but unfortunately the symbol is rather strange, and (as discussed in Section 4) it seems unlikely to gain wide adoption. (*Update*: This indeed proved to be the case, and Palais himself has since become a strong supporter of the arguments in this man-

Figure 3: The strange symbol for the circle constant from "π Is Wrong!".

ifesto.)

The Tau Manifesto is dedicated to the proposition that the proper response to "π is wrong" is "No, *really*." And the true circle constant deserves a proper name. As you may have guessed by now, *The Tau Manifesto* proposes that this name should be the Greek letter τ (tau):

$$\tau \equiv \frac{C}{r} = 6.283185307179586\ldots \qquad (3)$$

Throughout the rest of this manifesto, we will see that the *number* τ is the correct choice, and we will show through usage (Section 2 and Section 3) and by direct argumentation (Section 4) that the *letter* τ is a natural choice as well.

1.2   A powerful enemy

Before proceeding with the demonstration that τ is the natural choice for the circle constant, let us first acknowledge what we are up against—for there is a powerful conspiracy, centuries old, determined to propagate pro-π propaganda. Entire books are written extolling the virtues of π. (I mean, *books*!) And irrational devotion to π has spread even to the highest levels of geekdom; for example, on "Pi Day" 2010 Google *changed its logo* to honor π (Figure 4).

Meanwhile, some people memorize dozens, hundreds, even *thousands* of digits of this mystical number. What kind of sad sack memorizes even 40 digits of π (Figure 5)?[4]

---

[4]The video in Figure 5 (available at https://vimeo.com/12914981) is an excerpt

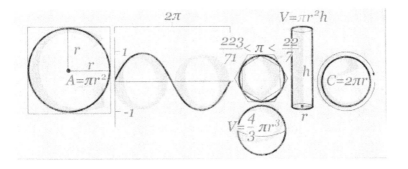

Figure 4: The Google logo on March 14 (3/14), 2010 ("Pi Day").

Figure 5: Matt Groening, incorrectly reciting π, says "Prove me wrong!"—so I do.

Truly, proponents of τ face a mighty opponent. And yet, we have a powerful ally — for the truth is on our side.

2   The number tau

We saw in Section 1.1 that the number τ can also be written as 2π. As noted in "π Is Wrong!", it is therefore of great interest to discover that the combination 2π occurs with astonishing frequency throughout mathematics. For example, consider integrals over all space in polar coordinates:

$$\int_0^{2\pi} \int_0^{\infty} f(r, \theta)\, r \, dr \, d\theta.$$

The upper limit of the θ integration is always 2π. The same factor appears in the definition of the Gaussian (normal) distribution,

$$\frac{1}{\sqrt{2\pi}\sigma} e^{-\frac{(x-\mu)^2}{2\sigma^2}},$$

and again in the Fourier transform,

$$f(x) = \int_{-\infty}^{\infty} F(k)\, e^{2\pi i k x} \, dk$$

$$F(k) = \int_{-\infty}^{\infty} f(x)\, e^{-2\pi i k x} \, dx.$$

It recurs in Cauchy's integral formula,

$$f(a) = \frac{1}{2\pi i} \oint_{\gamma} \frac{f(z)}{z - a} \, dz,$$

in the nth roots of unity,

$$z^n = 1 \Rightarrow z = e^{2\pi i/n},$$

from a lecture given by Dr. Sarah Greenwald, a professor of mathematics at Appalachian State University. Dr. Greenwald uses math references from *The Simpsons* and *Futurama* to engage her students' interest and to help them get over their math anxiety. She is also the maintainer of the *Futurama* Math Page.

and in the values of the Riemann zeta function for positive even integers:[5]

$$\zeta(2n) = \sum_{k=1}^{\infty} \frac{1}{k^{2n}}$$

$$= \frac{|B_{2n}|}{2(2n)!} (2\pi)^{2n}, \qquad n = 1, 2, 3, \ldots$$

These formulas are not cherry-picked—crack open your favorite physics or mathematics text and try it yourself. There are many more examples, and the conclusion is clear: there is something special about 2π.

To get to the bottom of this mystery, we must return to first principles by considering the nature of circles, and especially the nature of *angles*. Although it's likely that much of this material will be familiar, it pays to revisit it, for this is where the true understanding of τ begins.

2.1 Circles and angles

There is an intimate relationship between circles and angles, as shown in Figure 6. Since the concentric circles in Figure 6 have different radii, the lines in the figure cut off different lengths of arc (or *arc lengths*), but the angle θ (theta) is the same in each case. In other words, the size of the angle does not depend on the radius of the circle used to define the arc. The principal task of angle measurement is to create a system that captures this radius-invariance.

Perhaps the most elementary angle system is *degrees*, which breaks a circle into 360 equal parts. One result of this system is the set of special angles (familiar to students of trigonometry) shown in Figure 7.

A more fundamental system of angle measure involves a direct comparison of the arc length s with the radius r. Although the lengths in Figure 6 differ, the arc length grows in proportion to the radius, so the *ratio* of the arc length to the radius is the same in each case:

$$s \propto r \Rightarrow \frac{s_1}{r_1} = \frac{s_2}{r_2}.$$

---

[5] Here B_n is the nth Bernoulli number.

7

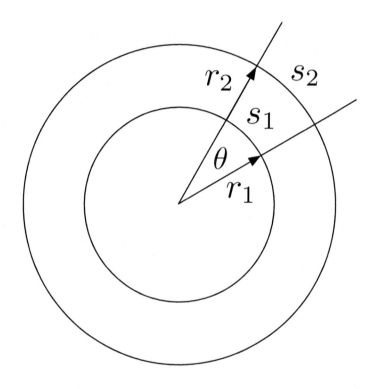

Figure 6: An angle θ with two concentric circles.

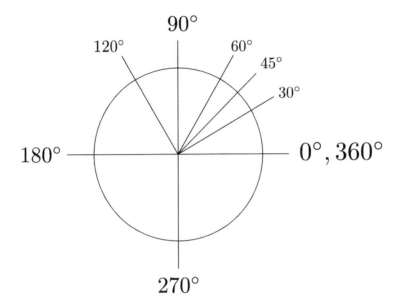

Figure 7: Some special angles, in degrees.

This suggests the following definition of *radian angle measure*:

$$\theta \equiv \frac{s}{r}.$$

(4)

This definition has the required property of radius-invariance, and since both s and r have units of length, radians are *dimensionless* by construction. The use of radian angle measure leads to succinct and elegant formulas throughout mathematics; for example, the usual formula for the derivative of $\sin\theta$ is true only when θ is expressed in radians:

$$\frac{d}{d\theta}\sin\theta = \cos\theta. \qquad \text{(only in radians)}$$

Naturally, the special angles in Figure 7 can be expressed in radians, and when you took high-school trigonometry you probably memorized the special values shown in Figure 8. (I call this system of measure π-radians to emphasize that they are written in terms of π.)

Now, a moment's reflection shows that the so-called "special" angles are just particularly simple *rational fractions* of a full circle, as shown in Figure 9. This suggests revisiting Eq. (4), rewriting the arc length s in terms of the fraction f of the full circumference C, i.e., $s = fC$:

$$\theta = \frac{s}{r} = \frac{fC}{r} = f\left(\frac{C}{r}\right) \equiv f\tau.$$

(5)

Notice how naturally τ falls out of this analysis. If you are a believer in π, I fear that the resulting diagram of special angles (Figure 10) will shake your faith to its very core.

Although there are many other arguments in τ's favor, Figure 10 may be the most striking. We also see from Figure 10 the genius of Bob Palais' identification of the circle constant as "one turn": τ is the radian angle measure for one *turn* of a circle. Moreover, note that with τ there is *nothing to memorize*: a twelfth of a turn is $\tau/12$, an eighth of a turn is $\tau/8$, and so on. Using τ gives us the best of both worlds by combining conceptual clarity with all the concrete benefits of radians; the abstract meaning of, say, $\tau/12$ is obvious, but it is also just a number:

$$\text{a twelfth of a turn} = \frac{\tau}{12} \approx \frac{6.283185}{12}$$
$$= 0.5235988.$$

10

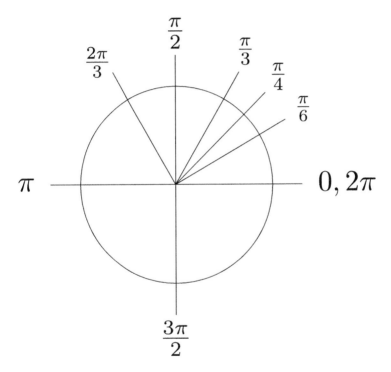

Figure 8: Some special angles, in π-radians.

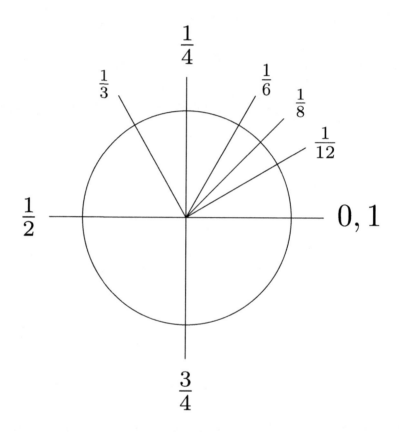

Figure 9: The "special" angles as fractions of a full circle.

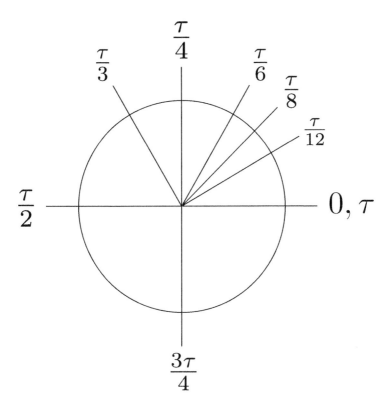

Figure 10: Some special angles, in radians.

Finally, by comparing Figure 8 with Figure 10, we see where those pesky factors of 2π come from: one turn of a circle is 1τ, but 2π. Numerically they are equal, but conceptually they are quite distinct.

2.1.1 The ramifications

The unnecessary factors of 2 arising from the use of π are annoying enough by themselves, but far more serious is their tendency to *cancel* when divided by any even number. The absurd results, such as a *half* π for a *quarter* turn, obscure the underlying relationship between angle measure and the circle constant. To those who maintain that it "doesn't matter" whether we use π or τ when teaching trigonometry, I simply ask you to view Figure 8, Figure 9, and Figure 10 through the eyes of a child. You will see that, from the perspective of a beginner, *using π instead of τ is a pedagogical disaster*.

2.2 The circle functions

Although radian angle measure provides some of the most compelling arguments for the true circle constant, it's worth comparing the virtues of π and τ in some other contexts as well. We begin by considering the important elementary functions $\sin\theta$ and $\cos\theta$. Known as the "circle functions" because they give the coordinates of a point on the *unit circle* (i.e., a circle with radius 1), sine and cosine are the fundamental functions of trigonometry (Figure 11).

Let's examine the graphs of the circle functions to better understand their behavior.[6] You'll notice from Figure 12 and Figure 13 that both functions are *periodic* with period T. As shown in Figure 12, the sine function $\sin\theta$ starts at zero, reaches a maximum at a quarter period, passes through zero at a half period, reaches a minimum at three-quarters of a period, and returns to zero after one full period. Meanwhile, the cosine function $\cos\theta$ starts at a maximum, has a minimum at a half period, and passes through zero at one-quarter and three-quarters of a period (Figure 13). For reference, both figures show the value of θ (in radians) at each special point.

---

[6]These graphs were produced with the help of Wolfram|Alpha.

14

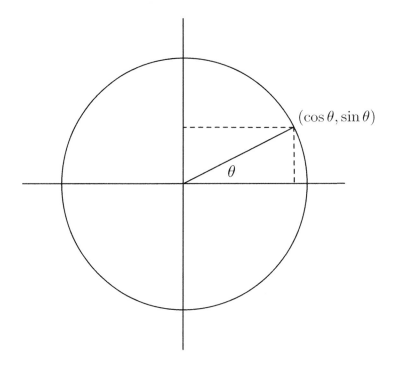

Figure 11: The circle functions are coordinates on the unit circle.

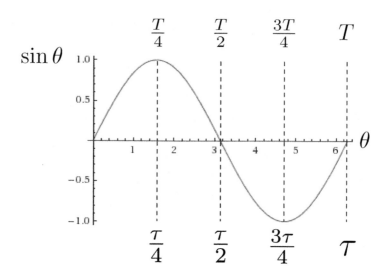

Figure 12: Important points for $\sin \theta$ in terms of the period T.

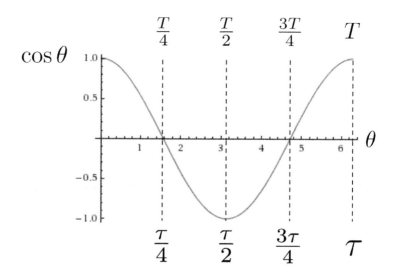

Figure 13: Important points for $\cos \theta$ in terms of the period T.

Of course, since sine and cosine both go through one full cycle during one turn of the circle, we have $T = \tau$; i.e., the circle functions have periods equal to the circle constant. As a result, the "special" values of θ are utterly natural: a quarter-period is $\tau/4$, a half-period is $\tau/2$, etc. In fact, when making Figure 12, at one point I found myself wondering about the numerical value of θ for the zero of the sine function. Since the zero occurs after half a period, and since $\tau \approx 6.28$, a quick mental calculation led to the following result:

$$\theta_{zero} = \frac{\tau}{2} \approx 3.14.$$

That's right: I was astonished to discover that *I had already forgotten that $\tau/2$ is sometimes called "π"*. Perhaps this even happened to you just now. Welcome to my world.

2.3 Euler's identity

I would be remiss in this manifesto not to address *Euler's identity*, sometimes called "the most beautiful equation in mathematics". This identity involves *complex exponentiation*, which is deeply connected both to the circle functions and to the geometry of the circle itself.

Depending on the route chosen, the following equation can either be proved as a theorem or taken as a definition; either way, it is quite remarkable:

$$e^{i\theta} = \cos\theta + i\sin\theta. \qquad \text{Euler's formula} \qquad (6)$$

Known as *Euler's formula* (after Leonhard Euler), this equation relates an exponential with imaginary argument to the circle functions sine and cosine and to the imaginary unit i. Although justifying Euler's formula is beyond the scope of this manifesto, its provenance is above suspicion, and its importance is beyond dispute.

Evaluating Eq. (6) at $\theta = \tau$ yields

$$e^{i\tau} = \cos\tau + i\sin\tau = 1 + 0i, \qquad (7)$$

which simplifies to *Euler's identity*:[7]

$$e^{i\tau} = 1. \qquad \text{Euler's identity (τ version)} \qquad (8)$$

---

[7] Here I'm implicitly defining Euler's identity to be *the complex exponential of the*

17

In words, Eq. (8) makes the following fundamental observation:

The complex exponential of the circle constant is unity.

Geometrically, multiplying by $e^{i\theta}$ corresponds to rotating a complex number by an angle θ in the complex plane, which suggests a second interpretation of Euler's identity:

A rotation by one turn is 1.

Since the number 1 is the multiplicative identity, the geometric meaning of $e^{i\tau} = 1$ is that rotating a point in the complex plane by one turn simply returns it to its original position.

As in the case of radian angle measure, we see how natural the association is between τ and one turn of a circle. Indeed, the identification of τ with "one turn" makes Euler's identity sound almost like a tautology.

2.3.1 Not the most beautiful equation

Of course, the traditional form of Euler's identity is written in terms of π instead of τ. To derive it, we start by evaluating Euler's formula at $\theta = \pi$, which yields

$$e^{i\pi} = \cos \pi + i \sin \pi = -1 + 0i \qquad (9)$$

and simplifes to

$$e^{i\pi} = -1. \qquad \text{Euler's identity (π version)} \qquad (10)$$

But that minus sign is so ugly that Eq. (10) is almost always rearranged immediately, giving the following "beautiful" equation:

$$e^{i\pi} + 1 = 0. \qquad \text{(rearranged)} \qquad (11)$$

---

circle constant, rather than defining it to be the complex exponential of any particular number. If we choose τ as the circle constant, we obtain the identity shown. As we'll see momentarily, this is not the traditional form of the identity, which of course involves π, but the version with τ is the most *mathematically* meaningful statement of the identity, so I believe it deserves the name.

At this point, the expositor usually makes some grandiose statement about how Eq. (11) relates $0, 1, i, e,$ and π—sometimes called the "five most important numbers in mathematics".

In this context, it's remarkable how many people complain that Eq. (8) relates only *four* of those five numbers. Fine:

$$e^{i\tau} = 1 + 0. \tag{12}$$

Indeed, we saw in Eq. (7) that there is actually a $0i$ already included (from $i \sin \tau$):

$$e^{i\tau} = 1 + 0i. \tag{13}$$

Eq. (13), *without* rearrangement, actually does relate the five most important numbers in mathematics: $0, 1, i, e,$ and τ.

2.3.2   Eulerian identities

Since you can add zero anywhere in any equation, the introduction of 0 in Eq. (12) is a somewhat tongue-in-cheek counterpoint to $e^{i\pi} + 1 = 0$, but the identity $e^{i\pi} = -1$ does have a more serious point to make. Let's see what happens when we rewrite it in terms of τ:

$$e^{i\tau/2} = -1.$$

Geometrically, this says that a rotation by half a turn is the same as multiplying by -1. And indeed this is the case: under a rotation of $\tau/2$ radians, the complex number $z = a + ib$ gets mapped to $-a - ib$, which is in fact just $-1 \cdot z$.

Written in terms of τ, we see that the "original" form of Euler's identity (Eq. (10)) has a transparent geometric meaning that it lacks when written in terms of π. (Of course, $e^{i\pi} = -1$ can be interpreted as a rotation by π radians, but the near-universal rearrangement to form $e^{i\pi} + 1 = 0$ shows how using π distracts from the identity's natural geometric meaning.) The quarter-angle identities have similar geometric interpretations: evaluating Eq. (6) at $\tau/4$ gives $e^{i\tau/4} = i$, which says that a quarter turn in the complex plane is the same as multiplication by i; similarly, $e^{i \cdot (3\tau/4)} = -i$ says that three-quarters of a turn is the same as multiplication by $-i$. A summary of these results, which we'll call *Eulerian identities*, appears in Table 1.

19

Rotation angle	Eulerian identity		
0	$e^{i \cdot 0}$	$=$	1
$\tau/4$	$e^{i\tau/4}$	$=$	i
$\tau/2$	$e^{i\tau/2}$	$=$	-1
$3\tau/4$	$e^{i \cdot (3\tau/4)}$	$=$	$-i$
τ	$e^{i\tau}$	$=$	1

Table 1: Eulerian identities for half, quarter, and full rotations.

We can take this analysis a step further by noting that, for any angle θ, $e^{i\theta}$ can be interpreted as a point lying on the unit circle in the complex plane. Since the complex plane identifies the horizontal axis with the real part of the number and the vertical axis with the imaginary part, Euler's formula tells us that $e^{i\theta}$ corresponds to the coordinates $(\cos\theta, \sin\theta)$. Plugging the values of the "special" angles from Figure 10 into Eq. (6) then gives the points shown in Table 2, and plotting these points in the complex plane yields Figure 14. A comparison of Figure 14 with Figure 10 quickly dispels any doubts about which choice of circle constant better reveals the relationship between Euler's formula and the geometry of the circle.

3   Circular area: the *coup de grâce*

If you arrived here as a π believer, you must by now be questioning your faith. τ is so natural, its meaning so transparent—is there no example where π shines through in all its radiant glory? A memory stirs—yes, there is such a formula—it is the formula for circular area! Behold:

$$A = \tfrac{1}{4}\pi D^2.$$

No, wait. The area formula is always written in terms of the *radius*, as follows:

$$A = \pi r^2.$$

We see here π, unadorned, in one of the most important equations in mathematics—a formula first proved by Archimedes himself. Order

Polar form	Rectangular form	Coordinates
$e^{i\theta}$	$\cos\theta + i\sin\theta$	$(\cos\theta, \sin\theta)$
$e^{i\cdot 0}$	1	$(1,0)$
$e^{i\tau/12}$	$\frac{\sqrt{3}}{2} + \frac{1}{2}i$	$(\frac{\sqrt{3}}{2}, \frac{1}{2})$
$e^{i\tau/8}$	$\frac{1}{\sqrt{2}} + \frac{1}{\sqrt{2}}i$	$(\frac{1}{\sqrt{2}}, \frac{1}{\sqrt{2}})$
$e^{i\tau/6}$	$\frac{1}{2} + \frac{\sqrt{3}}{2}i$	$(\frac{1}{2}, \frac{\sqrt{3}}{2})$
$e^{i\tau/4}$	i	$(0,1)$
$e^{i\tau/3}$	$-\frac{1}{2} + \frac{\sqrt{3}}{2}i$	$(-\frac{1}{2}, \frac{\sqrt{3}}{2})$
$e^{i\tau/2}$	-1	$(-1,0)$
$e^{i\cdot(3\tau/4)}$	$-i$	$(0,-1)$
$e^{i\tau}$	1	$(1,0)$

Table 2: Complex exponentials of the special angles from Figure 10.

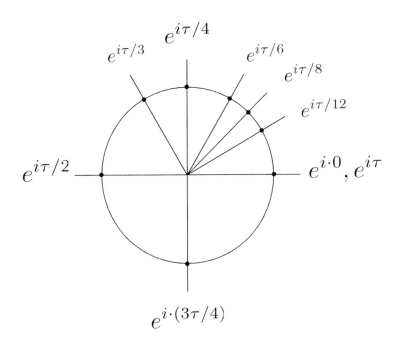

Figure 14: Complex exponentials of some special angles.

is restored! And yet, the name of this section sounds ominous… If this equation is π's crowning glory, how can it also be the *coup de grâce*?

3.1 Quadratic forms

Let us examine this putative paragon of π, $A = \pi r^2$. We notice that it involves the radius raised to the second power. This makes it a simple *quadratic form*. Such forms arise in many contexts; as a physicist, my favorite examples come from the elementary physics curriculum. We will now consider several in turn.

3.1.1 Falling in a uniform gravitational field

Galileo Galilei found that the velocity of an object falling in a uniform gravitational field is proportional to the time fallen:

$$v \propto t.$$

The constant of proportionality is the gravitational acceleration g:

$$v = gt.$$

Since velocity is the derivative of position, we can calculate the distance fallen by integration:[8]

$$y = \int v\,dt = \int_0^t gt\,dt = \tfrac{1}{2}gt^2.$$

3.1.2 Potential energy in a linear spring

Robert Hooke found that the external force required to stretch a spring is proportional to the distance stretched:

$$F \propto x.$$

---

[8]Technically, all the integrals should be definite, and the variable of integration should be different from the upper limit (as in $\int_0^t gt'\,dt'$, read as "the integral from zero to tee of gee tee prime dee tee prime"). These minor abuses of notation are common in physics and other less formal mathematical contexts such as we are considering here.

The constant of proportionality is the spring constant k:[9]

$$F = kx.$$

The potential energy in the spring is then equal to the work done by the external force:

$$U = \int F \, dx = \int_0^x kx \, dx = \tfrac{1}{2}kx^2.$$

3.1.3 Energy of motion

Isaac Newton found that the force on an object is proportional to its acceleration:

$$F \propto a.$$

The constant of proportionality is the mass m:

$$F = ma.$$

The energy of motion, or *kinetic energy*, is equal to the total work done in accelerating the mass to velocity v:

$$
\begin{aligned}
K = \int F \, dx = \int ma \, dx &= \int m\frac{dv}{dt} \, dx \\
&= \int m\frac{dx}{dt} \, dv \\
&= \int_0^v mv \, dv \\
&= \tfrac{1}{2}mv^2.
\end{aligned}
$$

3.2 A sense of foreboding

Having seen several examples of simple quadratic forms in physics, you may by now have a sense of foreboding as we return to the geometry of the circle. This feeling is justified.

---

[9]You may have seen this written as $F = -kx$. In this case, F refers to the force exerted by the *spring*. By Newton's third law, the external force discussed above is the *negative* of the spring force.

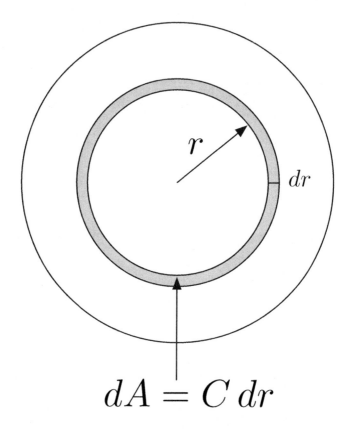

Figure 15: Breaking down a circle into rings.

Quantity	Symbol	Expression
Distance fallen	y	$\frac{1}{2}gt^2$
Spring energy	U	$\frac{1}{2}kx^2$
Kinetic energy	K	$\frac{1}{2}mv^2$
Circular area	A	$\frac{1}{2}\tau r^2$

Table 3: Some common quadratic forms.

As seen in Figure 15, the area of a circle can be calculated by breaking it down into circular rings of length C and width dr, where the area of each ring is $C\,dr$:

$$dA = C\,dr.$$

Now, the circumference of a circle is proportional to its radius:

$$C \propto r.$$

The constant of proportionality is τ:

$$C = \tau r.$$

The area of the circle is then the integral over all rings:

$$A = \int dA = \int_0^r C\,dr = \int_0^r \tau r\,dr = \tfrac{1}{2}\tau r^2.$$

If you were still a π partisan at the beginning of this section, your head has now exploded. For we see that even in this case, where π supposedly shines, in fact there is a missing factor of 2. Indeed, the original proof by Archimedes shows not that the area of a circle is πr^2, but that it is equal to the area of a right triangle with base C and height r. Applying the formula for triangular area then gives

$$A = \tfrac{1}{2}bh = \tfrac{1}{2}Cr = \tfrac{1}{2}\tau r^2.$$

There is simply no avoiding that factor of a half (Table 3).

3.2.1 Quod erat demonstrandum

We set out in this manifesto to show that τ is the true circle constant. Since the formula for circular area was just about the last, best argument that π had going for it, I'm going to go out on a limb here and say: Q.E.D.

4 Conflict and resistance

Despite the definitive demonstration of the superiority of τ, there are nevertheless many who oppose it, both as notation and as number. In this section, we address the concerns of those who accept the value but not the letter. We then rebut some of the many arguments marshaled against C/r itself, including the so-called "Pi Manifesto" that defends the primacy of π. In this context, we'll discuss the rather advanced subject of the volume of a hypersphere (Section 5.1), which augments and amplifies the arguments in Section 3 on circular area.

4.1 One turn

The true test of any notation is usage; having seen τ used throughout this manifesto, you may already be convinced that it serves its role well. But for a constant as fundamental as τ it would be nice to have some deeper reasons for our choice. Why not α, for example, or ω? What's so great about τ?

There are two main reasons to use τ for the circle constant. The first is that τ visually resembles π: after centuries of use, the association of π with the circle constant is unavoidable, and using τ feeds on this association instead of fighting it. (Indeed, the horizontal line in each letter suggests that we interpret the "legs" as *denominators*, so that π has two legs in its denominator, while τ has only one. Seen this way, the relationship $\tau = 2\pi$ is perfectly natural.)[10]

The second reason is that τ corresponds to one *turn* of a circle, and you may have noticed that "τ" and "turn" both start with a "t" sound. This was the original motivation for the choice of τ, and it is not a

---

[10]Thanks to *Tau Manifesto* reader Jim Porter for pointing out this interpretation.

coincidence: the root of the English word "turn" is the Greek word τόϱνος (tornos), which means "lathe". Using a math font for the first letter in τόϱνος then gives us: τ.

Since the original launch of *The Tau Manifesto*, I have learned that Peter Harremoës independently proposed using τ to "π Is Wrong!" author Bob Palais in 2010, John Fisher proposed τ in a Usenet post in 2004, and Joseph Lindenberg anticipated both the argument and the symbol more than twenty years before![11] Dr. Harremoës in particular has emphasized the importance of a point first made in Section 1.1: using τ gives the circle constant a *name*. Since τ is an ordinary Greek letter, people encountering it for the first time can pronounce it immediately. Moreover, unlike calling the circle constant a "turn", τ works well in both written and spoken contexts. For example, saying that a quarter circle has radian angle measure "one quarter turn" sounds great, but "turn over four radians" sounds awkward, and "the area of a circle is one-half turn r squared" sounds downright odd. Using τ, we can say "tau over four radians" and "the area of a circle is one-half tau r squared."

4.1.1 Ambiguous notation

Of course, with any new notation there is the potential for conflict with present usage. As noted in Section 1.1, "π Is Wrong!" avoids this problem by introducing a new symbol (Figure 3). There is precedent for this; for example, in the early days of quantum mechanics Max Planck introduced the constant h, which relates a light particle's energy to its frequency (through $E = h\nu$), but physicists soon realized that it is often more convenient to use $\hbar$ (read "h-bar")—where $\hbar$ is just h divided by… um… 2π—and this usage is now standard.

But getting a new symbol accepted is difficult: it has to be given a name, that name has to be popularized, and the symbol itself has to be added to word processing and typesetting systems. Moreover, promulgating a new symbol for 2π would require the cooperation of the academic mathematical community, which on the subject of π vs. τ

---

[11]Lindenberg has included both his original typewritten manuscript and a large number of other arguments at his site Tau Before It Was Cool (https://sites.google.com/site/taubeforeitwascool/).

has historically been apathetic at best and hostile at worst.[12] Using an existing symbol allows us to route around the mathematical establishment.[13]

Rather than advocating a new symbol, *The Tau Manifesto* opts for the use of an existing Greek letter. As a result, since τ is already used in some current contexts, we must address the conflicts with existing practice. Fortunately, there are surprisingly few common uses. Moreover, while τ is used for certain *specific* variables—e.g., *shear stress* in mechanical engineering, *torque* in rotational mechanics, and *proper time* in special and general relativity—there is no *universal* conflicting usage.[14] In those cases, we can either tolerate ambiguity or route around the few present conflicts by selectively changing notation, such as using N for torque,[15] τ_p for proper time, or even $\tau_\odot$ or τ for the circle constant itself.

Despite these arguments, potential usage conflicts have proven to be the greatest source of resistance to τ. Some correspondents have even flatly denied that τ (or, presumably, any other currently used symbol) could possibly overcome these issues. But scientists and engineers have a high tolerance for notational ambiguity, and claiming that τ-the-circle-constant can't coexist with other uses ignores considerable evidence to the contrary.

One example of easily tolerated ambiguity occurs in quantum mechanics, where we encounter the following formula for the *Bohr radius*, which (roughly speaking) is the "size" of a hydrogen atom in its

---

[12]I have been heartened, however, to see τ receive some significant support from mathematicians in the years following the publication of this manifesto. See, for example, "My Conversion to Tauism" by Stephen Abbott.

[13]Perhaps someday academic mathematicians will come to a consensus on a different symbol for the number 2π; if that ever happens, I reserve the right to support their proposed notation. But they have had over 300 years to fix this π problem, so I wouldn't hold my breath.

[14]The only possible exception to this is the *golden ratio*, which is often denoted by τ in Europe. But not only is there an existing common alternative to this notation—namely, the Greek letter φ (phi)—this usage shows that there is precedent for using τ to denote a fundamental mathematical constant.

[15]This alternative for torque is already in use; see, for example, *Classical Mechanics* (3rd edition) by Goldstein, Poole, and Safko, p. 2, and *Introduction to Electrodynamics* (4th edition) by David Griffiths, pp. 170–171.

lowest energy state (the *ground state*):

$$a_0 = \frac{\hbar^2}{me^2},\qquad (14)$$

where m is the mass of an electron and e is its charge. Meanwhile, the ground state itself is described by a quantity known as the *wavefunction*, which falls off exponentially with radius on a length scale set by the Bohr radius:

$$\psi(r) = N\,e^{-r/a_0},\qquad (15)$$

where N is a normalization constant.

Have you noticed the problem yet? Probably not, which is just the point. The "problem" is that the e in Eq. (14) and the e in Eq. (15) are *not the same* e — the first is the charge on an electron, while the second is the natural number (the base of natural logarithms). In fact, if we expand the factor of a_0 in the argument of the exponent in Eq. (15), we get

$$\psi(r) = N\,e^{-me^2 r/\hbar^2},$$

which has an e raised the power of something with e in it. It's even worse than it looks, because N itself contains e as well:

$$\psi(r) = \sqrt{\frac{1}{\pi a_0^3}}\,e^{-r/a_0} = \frac{m^{3/2}e^3}{\pi^{1/2}\hbar^3}\,e^{-me^2 r/\hbar^2}.$$

I have no doubt that if a separate notation for the natural number did not already exist, anyone proposing the letter e would be told it's impossible because of the conflicts with other uses. And yet, in practice no one ever has any problem with using e in both contexts above. There are many other examples, including situations where even π is used for two different things.[16] It's hard to see how using τ for multiple quantities is any different.

By the way, the π-pedants out there (and there have proven to be many) might note that hydrogen's ground-state wavefunction has a

---

[16]See, for instance, *An Introduction to Quantum Field Theory* by Peskin and Schroeder, where π is used to denote both the circle constant and a "conjugate momentum" on the very same page (p. 282).

factor of π:

$$\psi(r) = \sqrt{\frac{1}{\pi a_0^3}}\, e^{-r/a_0}.$$

At first glance, this appears to be more natural than the version with τ:

$$\psi(r) = \sqrt{\frac{2}{\tau a_0^3}}\, e^{-r/a_0}.$$

As usual, appearances are deceiving: the value of N comes from the product

$$\frac{1}{\sqrt{2\pi}}\frac{1}{\sqrt{2}}\frac{2}{a_0^{3/2}},$$

which shows that the circle constant enters the calculation through $1/\sqrt{2\pi}$, i.e., $1/\sqrt{\tau}$. As with the formula for circular area, the cancellation to leave a bare π is a coincidence.

4.2 The Pi Manifesto

Although most objections to τ come from scattered email correspondence and miscellaneous comments on the Web, there is also an organized resistance. In particular, after the publication of *The Tau Manifesto* in June 2010, a "Pi Manifesto" appeared to make the case for the traditional circle constant. This section and the two after it contain a rebuttal of its arguments.[17] Of necessity, this treatment is terser and more advanced than the rest of the manifesto, but even a cursory reading of what follows will give an impression of the weakness of the Pi Manifesto's case.

While we can certainly consider the appearance of the Pi Manifesto a good sign of continuing interest in this subject, it makes several false claims. For example, it says that the factor of 2π in the Gaussian (normal) distribution is a coincidence, and that it can more naturally be written as

$$\frac{1}{\sqrt{\pi}(\sqrt{2}\sigma)}e^{\frac{-x^2}{(\sqrt{2}\sigma)^2}}.$$

---

[17]The original Pi Manifesto has been removed (perhaps my rebuttal was a bit *too* effective?), so the link is now to an archived version (https://archive.md/VnJ2x).

This is wrong: the factor of 2π comes from squaring the unnormalized Gaussian distribution and switching to polar coordinates, which leads to a factor of 1 from the radial integral and a 2π from the angular integral. As in the case of circular area, the factor of π comes from $1/2 \times 2\pi$, not from π alone.

A related claim is that the gamma function evaluated at $1/2$ is more natural in terms of π:

$$\Gamma(\tfrac{1}{2}) = \sqrt{\pi},$$

where

$$\Gamma(p) = \int_0^\infty x^{p-1} e^{-x}\, dx. \tag{16}$$

But $\Gamma(\tfrac{1}{2})$ reduces to the same Gaussian integral as in the normal distribution (upon setting $u = x^{1/2}$), so the π in this case is really $1/2 \times 2\pi$ as well. Indeed, in many of the cases cited in the Pi Manifesto, the circle constant enters through an integral over all angles, i.e., as θ ranges from 0 to τ.

The Pi Manifesto also examines some formulas for regular n-sided polygons (or "n-gons"). For instance, it notes that the sum of the internal angles of an n-gon is given by

$$\sum_{i=1}^{n} \theta_i = (n-2)\pi.$$

This issue was dealt with in "π Is Wrong!", which notes the following: "The sum of the interior angles [of a triangle] is π, granted. But the sum of the *exterior* angles of *any* polygon, from which the sum of the interior angles can easily be derived, and which generalizes to the integral of the curvature of a simple closed curve, is 2π." In addition, the Pi Manifesto offers the formula for the area of an n-gon with unit radius (the distance from center to vertex),

$$A = n \sin\frac{\pi}{n} \cos\frac{\pi}{n},$$

calling it "clearly... another win for π." But using the double-angle identity $\sin\theta\cos\theta = \tfrac{1}{2}\sin 2\theta$ shows that this can be written as

$$A = n/2 \sin\frac{2\pi}{n},$$

31

which is just

$$A = \frac{1}{2}n \sin \frac{\tau}{n}. \tag{17}$$

In other words, the area of an n-gon has a natural factor of $1/2$. In fact, taking the limit of Eq. (17) as $n \to \infty$ (and applying L'Hôpital's rule) gives the area of a unit regular polygon with infinitely many sides, i.e., a unit circle:

$$\begin{aligned} A &= \lim_{n \to \infty} \frac{1}{2}n \sin \frac{\tau}{n} \\ &= \frac{1}{2} \lim_{n \to \infty} \frac{\sin \frac{\tau}{n}}{1/n} \\ &= \tfrac{1}{2}\tau. \end{aligned} \tag{18}$$

In this context, we should note that the Pi Manifesto makes much ado about π being the area of a unit disk, so that (for example) the area of a quarter (unit) circle is $\pi/4$. This, it is claimed, makes just as good a case for π as radian angle measure does for τ. Unfortunately for this argument, as noted in Section 3 and as seen again in Eq. (18), the factor of $1/2$ arises naturally in the context of circular area. Indeed, the formula for the area of a circular sector subtended by angle θ is

$$A(\theta) = \tfrac{1}{2}\theta r^2,$$

so there's no way to avoid the factor of $1/2$ in general. (We thus see that $A = \tfrac{1}{2}\tau r^2$ is simply the special case $\theta = \tau$.)

In short, the difference between angle measure and area isn't arbitrary. There is no natural factor of $1/2$ in the case of angle measure. In contrast, in the case of area the factor of $1/2$ arises through the integral of a linear function in association with a simple quadratic form. In fact, the case for π is even worse than it looks, as shown in the next section.

5   Getting to the bottom of pi and tau

I continue to be impressed with how rich this subject is, and my understanding of π and τ continued to evolve past the original Tau Day.

Most notably, on Half Tau Day 2012 I had an epiphany about *exactly* what is wrong with π. The argument hinges on an analysis of the surface area and volume of an n-dimensional ball, which makes clear that π as typically defined doesn't have any fundamental geometric significance.

The resulting section is more advanced than the rest of the manifesto and can be skipped without loss of continuity; if you find it confusing, I recommend proceeding directly to the conclusion in Section 6. But if you're up for a mathematical challenge, you are invited to proceed…

5.1   Surface area and volume of a hypersphere

We start our investigations with the generalization of a circle to arbitrary dimensions. This object, called a *hypersphere*, can be defined as follows. (For convenience, we assume that these spheres are centered on the origin.) A 0-sphere is the set of all points satisfying

$$x^2 = r^2,$$

which consists of the two points $\pm r$. These points form the boundary of a (closed) 1-ball, which is the set of all points satisfying

$$x^2 \leq r^2.$$

This is a line segment from $-r$ to r; equivalently, it is the closed interval $[-r, r]$.

A 1-sphere is a circle, which is the set of all points satisfying

$$x^2 + y^2 = r^2.$$

This figure forms the boundary of a 2-ball, which is the set of all points satisfying

$$x^2 + y^2 \leq r^2.$$

This is a closed disk of radius r; thus we see that the "area of a circle" is more properly defined as the area of a 2-ball. Similarly, a 2-sphere (also called simply a "sphere") is the set of all points satisfying

$$x^2 + y^2 + z^2 = r^2,$$

which is the boundary of a 3-ball, defined as the set of all points satisfying
$$x^2 + y^2 + z^2 \leq r^2.$$

The generalization to arbitrary n, although difficult to visualize for $n > 3$, is straightforward: an $(n - 1)$-sphere is the set of all points satisfying
$$\sum_{i=1}^{n} x_i^2 = r^2,$$
which forms the boundary of the corresponding n-ball, defined as the set of all points satisfying
$$\sum_{i=1}^{n} x_i^2 \leq r^2.$$

The "volume of a hypersphere" of dimension $n - 1$ is then defined as the volume $V_n(r)$ of the corresponding n-dimensional ball. It can be obtained by integrating the surface area $A_{n-1}(r)$ at each radius via $V_n(r) = \int A_{n-1}(r)\,dr$.

We will sometimes refer to A_{n-1} as the surface area of an n-dimensional ball, but strictly speaking it is the area of the *boundary* of the ball, which is just an $(n - 1)$-dimensional sphere. The subscripts on V_n and A_{n-1} are chosen so that they always agree with the dimensionality of the corresponding geometric object;[18] for example, the case $n = 2$ corresponds to a disk (dimensionality 2) and a circle (dimensionality $2 - 1 = 1$). Then V_2 is the "volume" of a 2-ball (i.e., the area of a disk, colloquially known as the "area of a circle"), and $A_{2-1} = A_1$ is the "surface area" of a 1-sphere (i.e., the circumference of a circle). When in doubt, simply recall that n always refers to the dimensionality of the *ball*, with $n - 1$ referring to the dimensionality of its boundary.

Now, The Pi Manifesto (discussed in Section 4.2) includes a formula for the volume of a unit n-ball as an argument in favor of π:

$$\frac{\sqrt{\pi}^n}{\Gamma(1 + \frac{n}{2})}, \tag{19}$$

---

[18]This choice of notation is fairly standard; see, e.g., the Wikipedia article on the volume of an n-ball.

where the gamma function is given by Eq. (16). Eq. (19) is a special case of the formula for general radius, which is also typically written in terms of π:

$$V_n(r) = \frac{\pi^{n/2} r^n}{\Gamma(1 + \frac{n}{2})}. \tag{20}$$

Because $V_n(r) = \int A_{n-1}(r)\, dr$, we have $A_{n-1}(r) = dV_n(r)/dr$, which means that the surface area can be written as follows:

$$A_{n-1}(r) = \frac{n \pi^{n/2} r^{n-1}}{\Gamma(1 + \frac{n}{2})}. \tag{21}$$

Rather than simply take these formulas at face value, let's see if we can untangle them to shed more light on the question of π vs. τ. We begin our analysis by noting that the apparent simplicity of the above formulas is an illusion: although the gamma function is notationally simple, in fact it is an integral over a semi-infinite domain (Eq. (16)), which is not a simple idea at all. Fortunately, the gamma function can be simplified in certain special cases. For example, when n is an integer, it is straightforward to show (using integration by parts) that

$$\Gamma(n) = (n-1)(n-2)\ldots 2 \cdot 1 = (n-1)!$$

Seen this way, $\Gamma(x)$ can be interpreted as a generalization of the factorial function to real-valued arguments.[19]

In the n-dimensional surface area and volume formulas, the argument of Γ is not necessarily an integer, but rather is $\left(1 + \frac{n}{2}\right)$, which is an integer when n is even and is a *half*-integer when n is odd. Taking this into account gives the following expression, which is adapted from a standard reference, Wolfram MathWorld, and as usual is written in terms of π:

$$A_{n-1}(r) = \begin{cases} \dfrac{2\pi^{n/2}\, r^{n-1}}{(\frac{1}{2}n - 1)!} & n \text{ even;} \\[3mm] \dfrac{2^{(n+1)/2}\pi^{(n-1)/2}\, r^{n-1}}{(n-2)!!} & n \text{ odd.} \end{cases} \tag{22}$$

---

[19] Indeed, the generalization to complex-valued arguments is straightforward: just replace real x with complex z in Eq. (16).

(Here we write A_{n-1} where MathWorld uses S_n.) Integrating with respect to r then gives

$$V_n(r) = \begin{cases} \dfrac{\pi^{n/2} r^n}{\left(\frac{n}{2}\right)!} & n \text{ even;} \\[3mm] \dfrac{2^{(n+1)/2} \pi^{(n-1)/2} r^n}{n!!} & n \text{ odd.} \end{cases} \tag{23}$$

Let's examine Eq. (23) in more detail. Notice first that MathWorld uses the *double factorial function* $n!!$—but, strangely, it uses it only in the *odd* case. (This is a hint of things to come.) The double factorial function, although rarely encountered in mathematics, is elementary: it's like the normal factorial function, but involves subtracting 2 at a time instead of 1, so that, e.g., $5!! = 5 \cdot 3 \cdot 1$ and $6!! = 6 \cdot 4 \cdot 2$. In general, we have

$$n!! = \begin{cases} n(n-2)\ldots 6 \cdot 4 \cdot 2 & n \text{ even;} \\[3mm] n(n-2)\ldots 5 \cdot 3 \cdot 1 & n \text{ odd.} \end{cases} \tag{24}$$

(By definition, $0!! = 1!! = 1$.) Note that Eq. (24) naturally divides into even and odd cases, making MathWorld's decision to use it only in the odd case still more mysterious.

To solve this mystery, we'll start by taking a closer look at the formula for odd n in Eq. (23):

$$\frac{2^{(n+1)/2} \pi^{(n-1)/2} r^n}{n!!}$$

Upon examining the expression

$$2^{(n+1)/2} \pi^{(n-1)/2},$$

we notice that it can be rewritten as

$$2(2\pi)^{(n-1)/2},$$

and here we recognize our old friend 2π.

36

Now let's look at the even case in Eq. (23). We noted above how strange it is to use the ordinary factorial in the even case but the double factorial in the odd case. Indeed, because the double factorial is already defined piecewise, if we unified the formulas by using $n!!$ in both cases we could pull it out as a common factor:

$$V_n(r) = \frac{1}{n!!} \times \begin{cases} \ldots & n \text{ even}; \\ \\ \ldots & n \text{ odd}. \end{cases}$$

So, is there any connection between the factorial and the double factorial? Yes—when n is even, the two are related by the following identity:

$$\left(\frac{n}{2}\right)! = \frac{n!!}{2^{n/2}} \qquad (n \text{ even}).$$

(This can be verified using mathematical induction.) Substituting this into the volume formula for even n in Eq. (23) then yields

$$\frac{2^{n/2} \pi^{n/2} r^n}{n!!},$$

which bears a striking resemblance to

$$\frac{(2\pi)^{n/2} r^n}{n!!},$$

and again we find a factor of 2π.

Putting these results together, we see that Eq. (23) can be rewritten as

$$V_n(r) = \begin{cases} \dfrac{(2\pi)^{n/2} r^n}{n!!} & n \text{ even}; \\ \\ \dfrac{2(2\pi)^{(n-1)/2} r^n}{n!!} & n \text{ odd} \end{cases} \qquad (25)$$

and Eq. (22) can be rewritten as

$$A_{n-1}(r) = \begin{cases} \dfrac{(2\pi)^{n/2} r^{n-1}}{(n-2)!!} & n \text{ even}; \\ \\ \dfrac{2(2\pi)^{(n-1)/2} r^{n-1}}{(n-2)!!} & n \text{ odd}. \end{cases} \qquad (26)$$

37

Making the substitution $\tau = 2\pi$ in Eq. (26) then yields

$$
A_{n-1}(r) = \begin{cases} \dfrac{\tau^{n/2}\, r^{n-1}}{(n-2)!!} & n \text{ even}; \\[3mm] \dfrac{2\tau^{(n-1)/2}\, r^{n-1}}{(n-2)!!} & n \text{ odd}. \end{cases}
$$

To unify the formulas further, we can use the *floor function* $\lfloor x \rfloor$, which is simply the largest integer less than or equal to x (equivalent to chopping off the fractional part, so that, e.g., $\lfloor 3.7 \rfloor = \lfloor 3.2 \rfloor = 3$). This gives

$$
A_{n-1}(r) = \begin{cases} \dfrac{\tau^{\lfloor \frac{n}{2} \rfloor}\, r^{n-1}}{(n-2)!!} & n \text{ even}; \\[3mm] \dfrac{2\tau^{\lfloor \frac{n}{2} \rfloor}\, r^{n-1}}{(n-2)!!} & n \text{ odd}, \end{cases}
$$

which allows us to write the formula as follows:

$$
A_{n-1}(r) = \frac{\tau^{\lfloor \frac{n}{2} \rfloor}\, r^{n-1}}{(n-2)!!} \times \begin{cases} 1 & n \text{ even}; \\[2mm] 2 & n \text{ odd}. \end{cases} \tag{27}
$$

Integrating Eq. (27) with respect to r then yields

$$
V_n(r) = \frac{\tau^{\lfloor \frac{n}{2} \rfloor}\, r^{n}}{n!!} \times \begin{cases} 1 & n \text{ even}; \\[2mm] 2 & n \text{ odd}. \end{cases} \tag{28}
$$

Note that, unlike the *faux* simplicity of Eq. (20), which hides a huge amount of complexity in the Γ function, Eq. (28) involves no fancy integrals—just the slightly exotic but nevertheless elementary floor and double-factorial functions.[20]

---

[20]Tau correspondent Jeff Cornell has pointed out that Eq. (27) and Eq. (28) can be further simplified by writing them in terms of the measure of a right angle, which he calls *lambda*: $\lambda \equiv \tau/4$. The resulting formulas effectively absorb the explicit

5.1.1 Recurrences

As seen in Eq. (27) and Eq. (28), the surface area and volume formulas divide naturally into two families, corresponding to even- and odd-dimensional spaces. This means that the surface area of a four-dimensional ball, $A_{4-1} = A_3$, is related to A_1 but not to A_2, while A_2 is related to A_0 but not to A_1 (and likewise for V_4 and V_2, etc.). How exactly are they related?

We can find the answer by deriving the *recurrence relations* between dimensions. In particular, let's divide the surface area of an n-dimensional ball by the surface area of an $(n-2)$-dimensional ball:

$$
\frac{A_{n-1}(r)}{A_{(n-2)-1}(r)} = \frac{\tau^{\lfloor \frac{n}{2} \rfloor}}{\tau^{\lfloor \frac{n-2}{2} \rfloor}} \frac{(n-2-2)!!}{(n-2)!!} \frac{r^{n-1}}{r^{n-3}}
$$
$$
= \frac{\tau}{n-2} r^2. \tag{29}
$$

Note that the different constants for even and odd n cancel out, thereby eliminating the dependence on parity. Similarly, for the ratio of the volumes we get this:

$$
\frac{V_n(r)}{V_{n-2}(r)} = \frac{\tau^{\lfloor \frac{n}{2} \rfloor}}{\tau^{\lfloor \frac{n-2}{2} \rfloor}} \frac{(n-2)!!}{n!!} \frac{r^n}{r^{n-2}}
$$
$$
= \frac{\tau}{n} r^2. \tag{30}
$$

---

dependence on parity into the floor function itself:

$$
A_{n-1}(r) = \frac{2^n \lambda^{\lfloor \frac{n}{2} \rfloor} r^{n-1}}{(n-2)!!}
$$

and

$$
V_n(r) = \frac{2^n \lambda^{\lfloor \frac{n}{2} \rfloor} r^n}{n!!}.
$$

To my knowledge, these are the most compact expressions of the spherical surface area and volume formulas. Their simplicity comes at the cost of a factor of 2^n, but this has a clear geometric meaning: a sphere in n dimensions divides naturally into 2^n congruent pieces, corresponding to the 2^n families of solutions to $\sum_{i=1}^{n} x_i^2 = r^2$ (one for each choice of $\pm x_i$). In two dimensions, they are the four quadrants; in three dimensions, they are the eight octants; and so on in higher dimensions. Nevertheless, because the dependence on parity is real and unavoidable (see, e.g., Figure 16), we will continue to write the formulas in terms of τ as in Eq. (27) and Eq. (28).

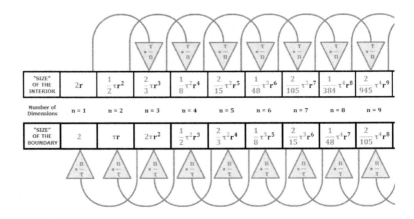

"SIZE" OF THE INTERIOR	$2r$	$\frac{1}{2}\tau r^2$	$\frac{2}{3}\tau r^3$	$\frac{1}{8}\tau^2 r^4$	$\frac{2}{15}\tau^2 r^5$	$\frac{1}{48}\tau^3 r^6$	$\frac{2}{105}\tau^3 r^7$	$\frac{1}{384}\tau^4 r^8$	$\frac{2}{945}\tau^4 r^9$
Number of Dimensions	$n=1$	$n=2$	$n=3$	$n=4$	$n=5$	$n=6$	$n=7$	$n=8$	$n=9$
"SIZE" OF THE BOUNDARY	2	τr	$2\tau r^2$	$\frac{1}{2}\tau^2 r^3$	$\frac{2}{3}\tau^2 r^4$	$\frac{1}{8}\tau^3 r^5$	$\frac{2}{15}\tau^3 r^6$	$\frac{1}{48}\tau^4 r^7$	$\frac{2}{105}\tau^4 r^8$

Figure 16: Surface area and volume recurrences.

We see from Eq. (29) and Eq. (30) that we can obtain the surface area and volume of an n-ball simply by multiplying the formula for an $(n-2)$-ball by r^2 (a factor required by dimensional analysis), dividing by $n-2$ or n, respectively, and multiplying by τ. As a result, τ provides the common thread tying together the two families of even and odd solutions, as illustrated by Joseph Lindenberg in Tau Before It Was Cool[21] (Figure 16).[22]

5.2 Three families of constants

Equipped with the tools developed in Section 5.1, we're now ready to get to the bottom of π and τ. To complete the excavation, we'll use Eq. (27) and Eq. (28) to define two families of constants, and then use the definition of π (Eq. (1)) to define a third, thereby revealing exactly what is wrong with π.

---

[21] Tau Before It Was Cool actually writes the recurrence in terms of 2π; the version shown in Figure 16 was created for me by special request. As always, I am most grateful to Joseph Lindenberg for his continuing generosity and support.

[22] Note that Figure 16 refers to the "size" of the *interior*; technically speaking, the interior is an open ball, whereas the volume in Eq. (28) is defined in terms of a closed ball, but the volumes of open and closed balls of the same radius are equal since the volume of the boundary is zero.

First, we'll define a family of "surface area constants" τ_{n-1} by dividing Eq. (27) by r^{n-1}, the power of r needed to yield a dimensionless constant for each value of n:

$$\tau_{n-1} \equiv \frac{A_{n-1}(r)}{r^{n-1}} = \frac{\tau^{\lfloor \frac{n}{2} \rfloor}}{(n-2)!!} \times \begin{cases} 1 & n \text{ even;} \\ 2 & n \text{ odd.} \end{cases} \tag{31}$$

Second, we'll define a family of "volume constants" v_n by dividing the volume formula Eq. (28) by r^n, again yielding a dimensionless constant for each value of n:

$$v_n \equiv \frac{V_n(r)}{r^n} = \frac{\tau^{\lfloor \frac{n}{2} \rfloor}}{n!!} \times \begin{cases} 1 & n \text{ even;} \\ 2 & n \text{ odd.} \end{cases} \tag{32}$$

With the two families of constants defined in Eq. (31) and Eq. (32), we can write the surface area and volume formulas (Eq. (27) and Eq. (28)) compactly as follows:

$$A_{n-1}(r) = \tau_{n-1}\, r^{n-1}$$

and

$$V_n(r) = v_n\, r^n.$$

Because of the relation $V_n(r) = \int A_{n-1}(r)\, dr$, we have the simple relationship

$$v_n = \frac{\tau_{n-1}}{n}.$$

Let us make some observations about these two families of constants. The family τ_{n-1} has an important geometric meaning: by setting $r = 1$ in Eq. (31), we see that each τ_{n-1} is the surface area of a unit $(n-1)$-sphere, which is also the angle measure of a full $(n-1)$-sphere. In particular, by writing $s_{n-1}(r)$ as the $(n-1)$-dimensional "arclength" equal to a fraction f of the full surface area $A_{n-1}(r)$, we have the exact analogue of Eq. (5) in n dimensions:

$$\theta_{n-1} \equiv \frac{s_{n-1}(r)}{r^{n-1}} = \frac{f A_{n-1}(r)}{r^{n-1}} = f\left(\frac{A_{n-1}(r)}{r^{n-1}}\right) = f\tau_{n-1}.$$

41

Here θ_{n-1} is simply the n-dimensional generalization of radian angle measure (where as usual n refers to the dimensionality of the corresponding ball), and we see that τ_{n-1} is the generalization of "one turn" to n dimensions. In the special case $n = 2$, we have the 1-sphere or circle constant $\tau_{2-1} = \tau_1 = \tau$, leading to the diagram shown in Figure 10.

Meanwhile, the v_n are the volumes of unit n-balls. In particular, v_2 is the area of a unit disk:

$$v_2 = \frac{\tau_1}{2} = \frac{\tau}{2}.$$

This shows that $v_2 = \tau/2 = 3.14159\ldots$ does have an independent geometric significance. Note, however, that *it has nothing to do with circumferences or diameters*. In other words, $\pi = C/D$ is not a member of the family v_n.

So, to which family of constants does π naturally belong? Let's rewrite Eq. (1) in terms more appropriate for generalization to higher dimensions:

$$\pi = \frac{C}{D} = \frac{A_1}{D^{2-1}}.$$

We thus see that π is naturally associated with surface areas divided by the power of the diameter necessary to yield a dimensionless constant. This suggests introducing a third family of constants π_{n-1}:

$$\pi_{n-1} \equiv \frac{A_{n-1}(r)}{D^{n-1}}. \tag{33}$$

We can express this in terms of the family τ_{n-1} by substituting $D = 2r$ in Eq. (33) and applying Eq. (31):

$$\pi_{n-1} = \frac{A_{n-1}(r)}{D^{n-1}} = \frac{A_{n-1}(r)}{(2r)^{n-1}} = \frac{A_{n-1}(r)}{2^{n-1}r^{n-1}} = \frac{\tau_{n-1}}{2^{n-1}}.$$

We are now finally in a position to understand exactly what is wrong with π. The principal geometric significance of $3.14159\ldots$ is that it is the area of a unit disk. But this number comes from evaluating $v_n = \tau_{n-1}/n$ when $n = 2$:

$$v_2 = \frac{\tau_1}{2} = \frac{\tau}{2}.$$

42

It's true that this happens to equal π_1:

$$\pi_1 = \pi = \frac{\tau_1}{2^{2-1}} = \frac{\tau}{2}.$$

But this equality is a coincidence: it occurs only because 2^{n-1} happens to equal n when $n = 2$ (that is, $2^{2-1} = 2$). In all higher dimensions, n and 2^{n-1} are distinct. In other words, *the geometric significance of π is the result of a mathematical pun*.

6   Conclusion

Over the years, I have heard many arguments against the wrongness of π and against the rightness of τ, so before concluding our discussion allow me to answer some of the most frequently asked questions.

6.1   Frequently Asked Questions

- **Are you serious?**
  Of course. I mean, I'm having fun with this, and the tone is occasionally lighthearted, but there is a serious purpose. Setting the circle constant equal to the circumference over the diameter is an awkward and confusing convention. Although I would love to see mathematicians change their ways, I'm not particularly worried about them; they can take care of themselves. It is the neophytes I am most worried about, for they take the brunt of the damage: as noted in Section 2.1, π is a pedagogical disaster. Try explaining to a twelve-year-old (or to a thirty-year-old) why the angle measure for an eighth of a circle—one slice of pizza—is $\pi/8$. Wait, I meant $\pi/4$. See what I mean? It's madness—sheer, unadulterated madness.

- **How can we switch from π to τ?**
  The next time you write something that uses the circle constant, simply say "For convenience, we set $\tau = 2\pi$", and then proceed as usual. (Of course, this might just prompt the question, "Why would you want to do that?", and I admit it would be nice to have a place to point them to. If only someone would write,

43

say, a *manifesto* on the subject...) The way to get people to start using τ is to start using it yourself.

- **Isn't it too late to switch? Wouldn't all the textbooks and math papers need to be rewritten?**
  No on both counts. It is true that some conventions, though unfortunate, are effectively irreversible. For example, Benjamin Franklin's choice for the signs of electric charges leads to the most familiar example of electric current (namely, free electrons in metals) being positive when the charge carriers are negative, and vice versa—thereby cursing beginning physics students with confusing negative signs ever since.[23] To change this convention *would* require rewriting all the textbooks (and burning the old ones) since it is impossible to tell at a glance which convention is being used. In contrast, while *redefining* π is effectively impossible, we can switch from π to τ on the fly by using the conversion

$$\pi \leftrightarrow \tfrac{1}{2}\tau.$$

  It's purely a matter of mechanical substitution, completely robust and indeed fully reversible. The switch from π to τ can therefore happen incrementally; unlike a redefinition, it need not happen all at once.

- **Won't using τ confuse people, especially students?**
  If you are smart enough to understand radian angle measure, you are smart enough to understand τ—and why τ is actually *less* confusing than π. Also, there is nothing intrinsically confusing about saying "Let $\tau = 2\pi$"; understood narrowly, it's just a simple substitution. Finally, we can embrace the situation as a teaching opportunity: the idea that π might be wrong is *interesting*, and students can engage with the material by converting the equations in their textbooks from π to τ to see for themselves which choice is better.

---

[23]The sign of the charge carriers couldn't be determined with the technology of Franklin's time, so this isn't his fault. It's just bad luck.

- **Does any of this really matter?**

  Of course it matters. *The circle constant is important.* People care enough about it to write entire books on the subject, to celebrate it on a particular day each year, and to memorize tens of thousands of its digits. I care enough to write a whole manifesto, and you care enough to read it. It's precisely because it *does* matter that it's hard to admit that the present convention is wrong. (I mean, how do you break it to Rajveer Meena, a world-record holder, that he just recited 70,000 digits of one half of the true circle constant?) Since the circle constant is important, it's important to get it right, and we have seen in this manifesto that the right number is τ. Although π is of great *historical* importance, the *mathematical* significance of π is that it is one-half τ.

- **Why did anyone ever use π in the first place?**

  The origins of π-the-number are probably lost in the mists of time. I suspect that the convention of using C/D instead of C/r arose simply because it is easier to *measure* the diameter of a circular object than it is to measure its radius. But that doesn't make it good mathematics, and I'm surprised that Archimedes, who famously approximated the circle constant, didn't realize that C/r is the more fundamental number. As notation, π was popularized around 300 years ago by Leonhard Euler, based on the work of William Jones. For example, in his hugely influential two-volume work *Introductio in analysin infinitorum*, Euler uses π to denote the *semicircumference* (half-circumference) of a unit circle or the measure of a $180°$ arc.[24] Unfortunately, Euler doesn't explain why he introduces this factor of $1/2$, though it may be related to the occasional importance of the semiperimeter of a polygon. In any case, he immediately notes that sine and cosine have periodicity 2π, so he was certainly in a position to

---

[24]"*...pro quo numero, brevitatis ergo, scribam π, ita ut sit π = Semicircumferentiae Circuli, cujus Radius = 1, seu π erit longitudo Arcus 180 graduum.*" "...for which number, because of brevity, I may write π, so that π may be the equal of the semicircumference of a circle, whose radius equals 1, or π will be the length of an arc of 180 degrees." Euler, Leonhard, *Introductio in analysin infinitorum* (1748), Volume 1, Chapter VIII, p. 93. https://scholarlycommons.pacific.edu/euler-works/101. Both definitions are equivalent to C/D since $D = 2$ when $r = 1$ and $180°$ is $\frac{1}{2} C/r$.

see that he was measuring angles in terms of *twice* the period of the circle functions, making his choice all the more perplexing. He almost got it right, though: somewhat incredibly, Euler actually used the symbol π to mean *both* $3.14\ldots$ and $6.28\ldots$ at different times![25] What a shame that he didn't standardize on the more convenient convention.

- **Why does this subject interest you?**

  First, as a truth-seeker I care about correctness of explanation. Second, as a teacher I care about clarity of exposition. Third, as a hacker I love a nice hack. Fourth, as a student of history and of human nature I find it fascinating that the absurdity of π was lying in plain sight for centuries before anyone seemed to notice. Moreover, many of the people who missed the true circle constant are among the most rational and intelligent people ever to live. What else might be staring us in the face, just waiting for us to discover it?

- **Are you like a crazy person?**

  That's really none of your business, but no. Like everyone, I do have my idiosyncrasies, but I am to all external appearances normal in practically every way. You would never guess from meeting me that, far from being an ordinary citizen, I am in fact a notorious mathematical propagandist.

- **But what about puns?**

  We come now to the final objection. I know, I know, "π in the sky" is so very clever. And yet, τ itself is pregnant with possibilities. τism tells us: it is not τ that is a piece of π, but π that is a piece of τ—one-half τ, to be exact. The identity $e^{i\tau} = 1$ says: *"Be one with the τ."* And though the observation that *"A rotation by one turn is 1"* may sound like a τ-tology, it is the true nature of the τ. As we contemplate this nature to seek the way of the τ, we must remember that τism is based on reason, not on faith: τists are never πous.

---

[25] For instance, in his 1727 *Essay Explaining the Properties of Air*, Euler writes: "*Sumatur pro ratione radii ad peripheriem,* I : π...", "It is taken for the ratio of the radius to the periphery [circumference], 1 : π..."

6.2 Embrace the tau

We have seen in *The Tau Manifesto* that the natural choice for the circle constant is the ratio of a circle's circumference not to its diameter, but to its radius. This number needs a name, and I hope you will join me in calling it τ:

$$\text{circle constant} = \tau \equiv \frac{C}{r}$$
$$= 6.283185307179586\ldots$$

The usage is natural, the motivation is clear, and the implications are profound. Plus, it comes with a really cool diagram (Figure 17). We see in Figure 17 a movement through *yang* ("light, white, moving up") to $\tau/2$ and a return through *yin* ("dark, black, moving down") back to τ.[26] Using π instead of τ is like having *yang* without *yin*.

6.28 Tau Day

The Tau Manifesto first launched on Tau Day: June 28 (6/28), 2010. Tau Day is a time to celebrate and rejoice in all things mathematical.[27] If you would like to receive updates about τ, including notifications about possible future Tau Day events, please join the *Tau Manifesto* mailing list at tauday.com.[28] And if you think that the circular baked goods on Pi Day are tasty, just wait—Tau Day has twice as much pi(e)!

6.283 Acknowledgments

I'd first like to thank Bob Palais for writing "π Is Wrong!". I don't remember how deep my suspicions about π ran before I encountered that article, but "π Is Wrong!" definitely opened my eyes, and every section of *The Tau Manifesto* owes it a debt of gratitude. I'd also like to thank Bob for his helpful comments on this manifesto, and especially for being such a good sport about it.

---

[26] The interpretations of yin and yang quoted here are from *Zen Yoga: A Path to Enlightenment through Breathing, Movement and Meditation* by Aaron Hoopes.

[27] Since 6 and 28 are the first two *perfect numbers*, 6/28 is actually a *perfect* day.

[28] A direct link is also available at https://tauday.com/join.

47

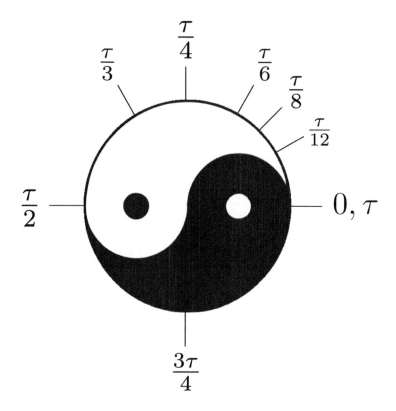

Figure 17: Followers of τism seek the way of the τ.

I've been thinking about *The Tau Manifesto* for a while now, and many of the ideas presented here were developed through conversations with my friend Sumit Daftuar. Sumit served as a sounding board and occasional Devil's advocate, and his insight as a teacher and as a mathematician influenced my thinking in many ways.

I have also received encouragement and helpful feedback from several readers. I'd like to thank Vi Hart and Michael Blake for their amazing τ-inspired videos, as well as Don "Blue" McConnell and Skona Brittain for helping make τ part of geek culture (through the time-in-τ iPhone app and the tau clock, respectively). The pleasing interpretation of the yin-yang symbol used in *The Tau Manifesto* is due to a suggestion by Peter Harremoës, who (as noted above) has the rare distinction of having independently proposed using τ for the circle constant. Another pre–*Tau Manifesto* τist, Joseph Lindenberg, has also been a staunch supporter, and his enthusiasm is much-appreciated.

I got several good suggestions from Christopher Olah, particularly regarding the geometric interpretation of Euler's identity, and Section 2.3.2 on Eulerian identities was inspired by an excellent suggestion from Timothy "Patashu" Stiles. Don Blaheta anticipated and inspired some of the material on hyperspheres, and John Kodegadulo put it together in a particularly clear and entertaining way. Then Jeff Cornell added a wonderful refinement with the introduction of $\lambda = \tau/4$, and Hjalmar Peters helped improve the exposition by persuading me to streamline the material on that subject.

I'd also like to acknowledge my appreciation for the volunteer translators who have made *The Tau Manifesto* available in so many different languages: Juan Guijarro Ferreiro (Spanish); Daniel Rosen and Alexis Drai (French); Andrea Laretto (Italian); Gustavo Chaves (Portuguese); Axel Scheithauer, Jonas Wagner, and Johannes Clemens Huber, with helpful notes from Caroline Steiblin (German); Aleksandr Alekseevich Adamov (Russian); and Daniel Li Qu (simplified Chinese).

Finally, I'd like to thank Wyatt Greene for his extraordinarily helpful feedback on a pre-launch draft of the manifesto; among other things, if you ever need someone to tell you that "pretty much all of the [now deleted] final section is total crap", Wyatt is your man.

49

6.2831 About the author

Michael Hartl is a physicist and entrepreneur. He is the author of over a dozen books, including *Learn Enough Python to Be Dangerous* and the *Ruby on Rails Tutorial*, and was cofounder and principal author at Learn Enough (acquired 2022). Previously, Michael taught theoretical and computational physics at the California Institute of Technology (Caltech), where he received a Lifetime Achievement Award for Excellence in Teaching and served as Caltech's editor for *The Feynman Lectures on Physics*. He is a graduate of Harvard College, has a Ph.D. in Physics from Caltech, and is an alumnus of the Y Combinator entrepreneur program.

Michael is ashamed to admit that he knows π to 50 decimal places—approximately 48 more than Matt Groening. To atone for this, he has memorized 52 decimal places of τ.

6.28318 Copyright

The Tau Manifesto. Copyright © 2010–2023 Michael Hartl. Please feel free to distribute *The Tau Manifesto* PDF[29] for educational purposes, and consider buying one or more copies of the print edition[30] for distribution to students and other interested parties.

6.283185 Dedication

The Tau Manifesto is dedicated to Harry "Woody" Woodworth, my eighth-grade science teacher. Although I gratefully received support from many teachers over the years, Woody believed in my potential to an extraordinary, even irrational (dare I say transcendental?) degree—confidently predicting that "someday they'll be teaching the 'Hartl theory' in schools." Given how many teachers have reached out indicating their support for and teaching of the material in *The Tau Manifesto*, I suppose in a sense Woody's prediction has now come true.

---

[29] https://tauday.com/pdf
[30] https://tauday.com/print

Printed in Great Britain
by Amazon

18767724R00031